SHALL WE SAVE THE EARTH?

Royston Fernandes

authorHOUSE®

AuthorHouse™ UK Ltd.
500 Avebury Boulevard
Central Milton Keynes, MK9 2BE
www.authorhouse.co.uk
Phone: 08001974150

First published by AuthorHouse 8/28/2007

ISBN: 978-1-4343-3514-2 (e)
ISBN: 978-1-4343-1049-1 (sc)

Printed in the United States of America
Bloomington, Indiana

This book is printed on acid-free paper.

FULL FORM OF CERTAIN ABBREVIATIONS
USED IN THIS BOOK

PSSS = PRESENT SYSTEM OF SELFISH SOCIETY

SWS = SOCIETY WITHOUT SELFISHNESS

PREFACE

The present system of selfish Society (PSSS in short) is so much selfish that in a span of 50 to 75 years this earth will be destroyed either by atom bombs or will be destroyed due to environmental catastrophe. It is estimated that fishes in ocean will be wiped off within the next 40 years. Recently a huge chunk of ice in North Canada simply claimed independence within a span of 2 hours and fell into the ocean. Thus man due to his selfish activities and due to lack of authority on the entire world as one entity is inflicting unimaginable damage to ecology and environment. The rich are only worried about becoming richer without regard to their poor compatriots. It is estimated that 2% of world's population is enjoying 50% of world's wealth. The words of Sunitha Williams, the astronut, during her space voayage clearly is a pointer to our selfish Society. **She said that from space she could see no boundaries between nations!** The boundaries and barriers are all man made by his selfish instincts. The PSSS because of its selfish goals and partisan policies and due to no global planning is going to destroy itself. Terrorism is showing its ugly teeth now and then and is threatening the world on religious sentiments.

A recent UN report says that 85 crore people on earth sleep every night without meals or food. Female foeticide is a common thing in developing and under developed countries like India, Pakistan etc. Mal-nutrition and under nutrition among children is common through out the developing and under developed countries.

Under such circumstances there is absolute necessity to establish Society without selfishness (SWS in short) to make this earth fit for living for the next generation. IF U WANT YOUR CHILDREN AND GRANDCHILDREN TO LIVE ON THIS EARTH IN A NORMAL ATMOSPHERE THEN PLEASE WAKE UP TO ESTABLISH SWS. How? Here is the constitution of the SWS. Please read it :

SOCIETY WITHOUT SELFISHNESS (SWS)
(The final chance for survival of this universe)

PREAMBLE:

Every EARTHIZEN (person living on this earth) WILL HAVE UNDIVIDED EQUAL SHARE ON IMMOVABLE PROPERTIES OF THIS EARTH. The basic idea for establishment of a Society without selfishness is to ensure equality in all respects to every earthizen. **The entire scheme of SWS is based on the simple motive of providing free food, clothing and shelter to every earthizen** while curbing all luxuries now enjoyed by selected few. All weapons including small arms will be destroyed. A common religion will be formulated and all persons and regions will be re-named. Towards this goal necessary basic corrections in the working of the present selfish Society are required. The Society without selfishness ensures maximum satisfaction with minimum resources. It is founded on the basic ideal of love everyone as much as you love yourself. It ensures restoration of environment to its natural form, which is being fast contaminated by present system. It ensures common amenities, religion, rights and duties to every earthizen. BASICS : LOVE EVERYONE WITHOUT SELFISHNESS.

1 . SWS SHALL FEED EVERYONE ON EARTH.
CURB ON MILITIARY AND WASTEFUL
EXPENDITURE AND TECHNOLOGICAL
ADAVANCES SHALL MAKE IT ACHIEVE THIS.

1
WHAT ARE THE CAUSES FOR FAILURE OF PRESENT SYSTEM OF SELFISH SOCIETY ?

The main cause for failure for the present system of selfish Society (PSSS) is its base. It is based on selfishness. **Everyone including Governments and religious leaders act and take decisions based on selfishness.** It gives scant respect to individuals. It respects only the successful. It practices slavery, treachery, corruption, murder, genocide, ethnic cleansing, monopoly, competition, terrorism, wars, drug abuse, regularisation of irregular activities, adultery, abortion, drunkenness, drug addiction, smoking and so on which are all the ugly faces of SELFISHNESS. (Called evils in future) About 75% of earthizens are below poverty line and fight a losing battle for survival. It is estimated that 50% of world's wealth is in the hands of just 2% people. They are the victims of slavery and almost all the evils mentioned above. Religions and self help groups render some help but the fact that poverty and hunger is predominant only shows that selfishness is getting upper hand. The PSSS is trying to satisfy the unlimited wants of a few selected successful persons at the cost of limited resources of this mother EARTH and in the process it pushes more and more people below the poverty line. They have evolved a system of Democracy stating that it gives right to the people. But unfortunately this Democracy has benefited the rulers and not the ruled. The political leaders throughout the world have practised every possible evils mentioned above. The present justice system has ensured their safety on the dogma no punishment without witness. Witnesses are suppressed and the guilty are ensured safe passage. Majority has lost faith in the system but unfortunately they are helpless in the absence of an alternative. COMPLEXITY IS THE ORDER OF THE DAY. The new system of Society without selfishness will ensure to eradicate these anomalies. SIMPLICTY WILL BE THE WATCHWORD.

Published: Saturday, April 7, 2007 Herald Net

U.N. climate report grim
Scientists say politics kept the final report on global warming from being even bleaker.

By Alan Zarembo and Thomas Maugh Ii
Los Angeles Times

A new global warming report issued Friday by the United Nations paints a near-apocalyptic vision of Earth's future: more than a billion people in need of water, extreme food shortages in Africa, a planetary landscape ravaged by floods and millions of species sentenced to extinction.

But despite the harshness of its vision, scientists who said government bureaucrats seeking to deflect calls for action watered down its findings at the last minute quickly criticised the report.

"The science got hijacked by the political bureaucrats at the late stage of the game," said John Walsh, a climate expert at the University of Alaska, Fairbanks, who helped author a chapter on the Polar Regions.

Even in its softened form, the report outlined a range of devastating effects that will strike all regions of the world and all levels of society. Those without resources to adapt to the changes will suffer the greatest impact, according to the study from the UN's Intergovernmental Panel on Climate Change.

2. SWS CANCELS ALL BOUNDARIES WITHIN A COUNTRY AND FINALLY BOUNDARIES BETWEEN COUNTRIES. THERE WILL BE ONLY ONE COUNTRY FINALLY CALLED 'EARTH'.

"It's the poorest of the poor in the world, and this includes poor people even in prosperous societies, who are going to be the worst hit," said Rajendra Pachauri, chairman of the IPCC, which released the report in Brussels, Belgium, on Friday.

The report is the second issued this year by the United Nations, which marshalled more than 2,500 scientists to give their best predictions of the consequences of a few degrees increase in temperature.

The first report, released in February, characterised global warming as a runaway train that is irreversible, but that can be moderated by societal changes. That report said, with more than 90 percent confidence that the warming is caused by humans, and its conclusions were widely accepted because of the years of accumulated scientific data supporting it.

In contrast, the second report was more controversial because it tackled the more uncertain issues of the precise effects of warming and the ability of humans to adapt to it.

"When you put people into the equation, people who can adapt and respond and change their behavior, it adds another layer of complication," said Gary Yohe, an economist at Wesleyan University who helped author the report.

The report, in a sense, is a more focused indictment of the world's biggest polluters - the industrialized nations - and a more specific identification of the victims.

3.   ONE CANNOT PUT BARRICADES OR
ENFORCE BARRIERS BETWEEN HUMAN
BEINGS. EARTH BELONGS TO ALL HUMAN
BEINGS.

The last-minute negotiations led to deleting timelines for future events and scaling back the degree of confidence in some projections. Both actions will ease the pressure on industrialized nations to reduce their emissions of carbon dioxide and other greenhouse gases that are gradually warming the planet.

Several scientists vowed afterward that they would never participate in the process again because of the political interference.

"Once is enough," said Walsh, who was not present during the negotiations in Brussels but kept abreast of developments with e-mails from colleagues. "I was receiving hourly reports that grew increasingly frustrated."

The report paints a bleak picture of the future, noting that the early signs of warming already are here:

* Spring is arriving earlier, with plants blooming weeks ahead of schedule.

* In North America, snowpack in the West will decline, causing more floods in the winter and reduced flows in the summer, increasing competition for water for agriculture and municipal use. California agriculture will be decimated by the loss of water for irrigation, experts have previously said. Water will come more often around the world in its least welcome forms: storms and floods.

* In the mountains, the runoff begins earlier in the year, shrinking glaciers in the Alps, the Himalayas and the Andes.

4. SWS WILL ESTABLISH FOR THE ENTIRE WORLD ONE SELFISHLESS GOVERNMENT WHERE LOVE DOMINATES ALL DECISIONS.

* Habitats for plants and animals, both on land and in the oceans, are shifting toward the poles.

* Nineteen of the 20 hottest years on record have occurred since 1980, according to previous studies. The report said more frequent and more intense heat waves are "very likely" in the future.

In some places, warming might seem like a good thing, at first. For example, worldwide food production is expected to increase with the first few degrees of temperature rise. For a time, an expanded fertile zone in the higher latitudes could offset losses in the tropics. But at a certain point, as drought conditions spread, crops everywhere will suffer.

* By mid-century, temperature rise and drying soil will replace tropical forests with savannas in Brazil's eastern Amazonia, the report predicts.

* Rising temperature will reconfigure coastlines around the world, as the oceans rise and seawater surges over land. The tiny islands of the South Pacific and the Asian deltas will be overwhelmed by storm surges as sea levels rise.

* In the Andes and the Himalayas, melting glaciers will unleash floods and rock avalanches. But within a few decades, as the glaciers and snowpack decline, streams will dwindle, cutting off the main water supply to more than one-sixth of the world's population.

* Africa will suffer the most extreme effects, with a quarter of a billion people losing most of their water

5.   SWS SHALL WORSHIP ONLY ONE GOD.
SWS CANNOT BE DEVIDED ON BASIS
OF RELIGION. SWS RELIGION WILL BE
RELIGION OF LOVE WITH ONLY ONE GOD.
HIS NAME WILL BE GOD FOR ALL PURPOSES.

supplies, the report said. Food production will fall by half in many countries and governments will have to spend 10 percent of their budgets or more to adapt to climate changes, the report said.

* At least 30 percent of the world's species will disappear if temperatures rise 3.6 degrees Fahrenheit above the average levels of the 1980s and 1990s, the report said.

"Don't be poor in a hot country, don't live in hurricane alley, watch out about being on the coasts or in the Arctic, and it's a bad idea to be on a high mountain," said Stephen Schneider of Stanford University, one of the scientists who contributed to the report.

The Bush administration quickly made it clear that it would not be stampeded by the report into taking part in the United Nations' Kyoto Protocol, which seeks to limit emissions of carbon dioxide. The U.S. withdrew from the protocol in 2001, saying it was too expensive and did not impose enough controls on developing nations.

"Each nation sort of defines their regulatory objectives in different ways to achieve the greenhouse reduction outcome that they seek," said Jim Connaughton, chairman of the White House Council on Environmental Quality, during a teleconference Friday from Brussels.

Sharon Hays, associate director of the White House Office of Science and Technology, noted in the same teleconference that "not all projected impacts are negative."

6.   NO MORE MURDERS, BOMB EXPLOSIONS, WARS, CONFLICTS, SUICIDES, SUICIDE BOMBINGS, ATTACKS, HIJACKING, KIDNAPPING. WHY? BECAUSE SWS HAS NO MONEY TO EARN FROM THESE ABUSES.

Initially, the warming will increase agricultural output in the mid-latitudes and in northern regions.

Other governments, such as China, Russia and Saudi Arabia, had already expressed their displeasure with parts of the report by demanding changes - some of them seemingly minor in the grand scheme of climate change.

Panel member Yohe said that China and Saudi Arabia, for example, objected to a sentence that stated "very high confidence" that many natural systems are already being affected by regional climate changes, arguing that "very" should be removed.

After a long deadlock, U.S. delegates brokered a compromise that removed the reference to confidence levels.

The U.S. delegation opposed a section that said parts of North America could suffer "severe" economic damage from climate change.

THE ABOVE REPORT SHOWS HOW RICH NATIONS ARE BLOCKING ATTEMPTS BY SCIENTISTS TO SAVE THE EARTH. THE GREATEST SUFFERERS OF CLIMATE CHANGE WILL BE POOR PEOPLE. THEREFORE WE SHOULD ERADICATE THE PSSS. PSSS HAS NO SOLUTIONS TO THE ABOVE CLIMATE REPORT. IT CANNOT THINK OF A DEVELOPMENTAL MODEL WHEREIN IT CAN BAN USE OF CRUDE OIL PRODUCTS WHICH ARE RESPONSIBLE FOR CARBON DIOXIDE EMISSION. THEREFORE IT

WILL SUPPRESS ANY VOICE RAISED AGAINST ENVIRONMENTAL POLLUTION. THE VOICE OF THE SCIENTISTS IS SUPPRESSED. PSSS IS WORRIED ABOUT ONLY THE PRESENT GENERATION. IT IS NOT BOTHERED ABOUT THE FUTURE GENERATION.

2
ANIMALS ARE FAR BETTER THAN MAN

Animals take share of this nature only to the extent of their daily requirement whereas the selfish MAN wants not only for today and for himself but wants to create wealth not only for himself but for his great grandchildren. The wants of animals are limited whereas wants of human beings are unlimited. Man does not respect any laws and rules either of this Nature or of God. He wants his wants to be satisfied at any cost giving scant respect to his fellow earthizens, nature or God.

ALL CHILDREN BORN ARE EQUAL IN THEIR BEHAVIOUR WHETHER THEY ARE BORN TO A RICH MAN OR A POOR MAN

Children speak the truth when it comes to equality. The behaviour of children whether born to a rich man or poor man is all equal. It is only the PSSS which when the child grows teaches the inequality and creates the have and have nots.

WOMEN ARE RESPONSIBLE FOR UPBRINGING BOY CHILD

All men should remember that their mother brought them up. Thus women who should have dominated the affairs of this Earth are unfortunately sidelined and harassed by men when they grow up. Men when they rape a girl should remember they are doing it against his own sister in her mother's resemblance.

7.   *SWS GURANTEES LIFE UNTO NATURAL DEATH.*

3
HOW THE NEW SYSTEM WILL WORK?

AT THE OUTSET IT IS MADE CLEAR DURING THE TRANSITION PERIOD the members are directed to stay in the PSSS and **obey all its laws and regulations. There will be not any revolt.** The transition period should not be violent. The members are required to stick on to basics. Otherwise the SWS will be derailed. Our preamble does not recognise violence or revolt.

The SWS will be established in 3 phases. Phase 1- Transition Phase 2- Universalisation Phase 3- Completion.

Phase 1 The Transition phase is divided into 5 steps:

1.Step 1- In this step like minded people will organise to develop and register regional SWS groups. During this step members will obey all the laws and rules of the PSSS. These groups will register and assign certain sequential numbers, which will become their names once the PSSS is established. These groups are required to enrol has political entity by name SWS. They shall propagate the principles of new system and they are required to be patient during their interaction without showing any signs of anger or violence. The members will continue to live in the PSSS without any obstacles to the existing system.The SWS will fight elections

in Democratic countries and its success in elections is rest assured because majority of the earthizens are victims of the PSSS.

2. Step 2 -Establishment of Governments:- Once SWS takes power in democratic countries it will not immediately establish SWS. Transition should not be violent. The following initial corrections will be made to PSSS: -

a) the salaries of all the employees working in Government and semi-Government organisations will be equalised. The employees will be taught that every human being is equal in hunger and naturally everyone requires equal food, clothing and shelter. The non-government establishments will soon follow path. Again while initiating this step every care should be taken to ensure the basic LOVE EVERYONE. The law equalising salary will try to educate the people and if any violence is resorted by employees negotiation and not coercion should deal the same. This step is most crucial for the establishment of SWS throughout the world. Care should be taken to implement this step with great caution as supporter of PSSS may resort to provocative violence. There should not be major change in the laws and rules until all the employees including menial labourers are ensured equal salary. The success of this step will

8.   NO AMASSMENT OF WEALTH. NO FEAR OF TOMORROW. SWS SHALL PROVIDE DAILY BREAD WITH LOVE ATTACHED.

be primary to the establishment of SWS throughout the world.

b)   b) EDUCATION: People will be educated that selfishness was the root cause for all the evils. They will be educated that equality in all respects to every earthizen is essential for eradicating selfishness. They should be educated and prepared to the next step of nationalisation of property to ensure equality in shelter to everyone.

c)   c) Religion: The religious leaders will be requested and persuaded to sacrifice their differences and devise a common religion. They will be enlightened that different religious ideologies have caused damage to everyone destroying peace. Peace and love are basics of all religions. **Naturally a common religion can be formulated to ensure peace and love among all**. Meanwhile people will be educated to sacrifice their different religious practices and ensure equality in religious thinking. God is one and hence naturally it is not correct to carry on different faiths, which has caused so much hatred and violence in the PSSS.

d)   d) RENAMING PLACE AND PERSONS: - This is essential to eradicate the stigma attached to names, which has caused caste system and divided the Society. Hitler can be best example for practising genocide for the sake of name. The people will be

9.   *WE CONTACTED VARIOUS GODS AND FOUND THAT HE IS ONLY ONE. HIS NAME IS GOD AND GOD ALONE.*

educated and mind prepared to change their names and the names of their places. They will be taught that hence forth a person will not more equal because of his name or name of the region from which he is hailing.

3.STEP-3 ESTABLISHING SWS IN PURE FORM: This step is most fundamental step in establishing SWS. The various actions are enumerated as follows:

a) **Nationalising private property**: All immovable properties will be nationalised. Property rights will be restricted to movable properties for the time being. Community centres on the principles of co-operation will be established. In this community centre people will be listed and arrangements will be made to ensure food, clothing and shelter. Movable properties will be nationalised in course of time.

b) b) **Re-naming of persons on the basis of alpha-numerical marks**: All persons attached to a community will be re-named based on a distinguishing alpha-numerical mark which will remove the stigma attached to a person of religion, caste and region.

c) c) **Common religion:** Religion in itself is good. But the variety in religions has created such a deep havoc in PSSS that even GOD has become a contentious entity. A common religion based on

10. **RELIGIOUS LEADERS AND POLITICIANS YOU FOUL PLAY IS DETECTED. PLEASE DO NOT INTERFERE IN THE ESTABLISHMENT OF SWS.**

the principle that GOD is one and should be given all respects within one mind and heart and not by public demonstration will be established. There will be no public demonstration of religious sentiments. Anyway the new system ensures that no sins are committed and naturally we do not need agents to pray for forgiveness with GOD for the sins.

4. STEP-4 UNIVERLISATION: - This last step ensures SWS throughout the world. The following actions are contemplated:

a) **Re-naming regions and countries**: This is essential to remove the stigma attached to a name of a region or a country. The new names will be alph-numerical.

b) b) **The entire world will be declared as one entity** and kings and despots and even certain puppet democracies, which have so far not fallen in line with SWS, will be persuaded to join the main stream. If they fail forceful acquisition is the only and the last violent action that will be adopted on this earth. Once the entire world is brought under SWS then no more violence will be practised anywhere in the world.

c) c) All weapons including small arms will be destroyed. Weapons will be STATE property to be

11.   FOOD ON EARTH BELONGS TO EVERY
EARTHIZEN. SWS SHALL ONLY DISTRIBUTE
IT EQUALLY.

used not against humans but in case of necessity against violent animals.

d) d) Researchers will be asked to ensure the best possible food for every earthizen with least pressure on nature.

e) e) Education is compulsory and free. English will be adopted as international language and every earthizen is required to study it. **The entire world will have English as official language.**

f) f) People will be required to lead a disciplined and life of sacrifice for the betterment of every one.

g) g) **The main industry will be agriculture** and the land available will be put to maximum use with traditional methods augmented by best possible non-environmental pollutant technology.

h) h) Marriages will be between two loved couples. The boy and girl are required to understand each other before Marriage and once married they cannot divorce. Hence only love marriages will be encouraged. Arranged marriages will be discouraged. Once married the couples will remain together till their separation due to death. Widows below the age of 35 for men and 30 for women will be permitted to re-marry if they wish to marry.

12.   LET US MAKE OUR CHILDREN, GRANDCHILDREN AND THEIR CHILDREN AND SO ON INHERIT A SAFE AND PURE ENVIRONMENT.

i) i) **All earthizens will enjoy maximum freedom in all walks of life**; however, they should ensure their freedom will not endanger SWS.

j) j) Movement of people will be restricted to the community to reduce unnecessary pollution and pressure on meagre natural resources.

k) k) **Industries will only ensure equal food, clothing and shelter**. All industries, which are not relevant in SWS, will be discarded. Prominence will be given on protecting nature of its virgin form.

l) l) Hospitals will provide free medicine to everyone. There will only be special wards!

m) m) State will ensure that all elderly, physically handicapped and insane persons will receive the same treatment as an EARTHIZEN. There will be special community centres to look after these people.

n) There will not be unemployment in the new system. Every healthy adult other than students will work as per his qualification. State will ensure that human resources are fully utilised without any scope for laziness. Education will ensure that people will work more for himself but for the Society.

o) o) **All forms of drugs like liquor, beer, tobacco and its products will be banned.**

p) **PUNISHMENTS UNDER SWS:** People not ready to accept SWS, lazy people, violent people

13. SWS PLANS FOR 100 YEARS AT A STRETCH AND PLANS AT THE LEVEL OF EARTH AND NOT AT THE LEVEL OF A NATION.

will be dealt in the following manner: They will be transferred to secluded community centres where they will be educated and persuaded to accept SWS. Once they fall in line they will be transferred back to their parent Community centre. There will not be capital punishments or punishment of any violent nature. The SWS ensures no crime because of its ideals. You may not find the necessity of stealing or robbing because you cannot enjoy property rights. The system ensure you free food, clothing and shelter and naturally when there are no other luxuries that can be enjoyed why you practice any violence or wicked act.

EVOLUTION OF PSSS

The PSSS evolved somewhere when man realised that he can make others work for him. Meaning to say the day he became lazy. Then he realised that selfishness can bring him happiness. But he failed to realise that even his brethren can practice selfishness. This created conflicts. He created religion to regulate selfishness. He preached and not practised.

Christianity became famous in masses because it dealt a blow to selfishness. Jesus Christ sacrificed his life on the crucifix. In the early 300 years of Christianity people realised the sacrifice of Jesus and they started uniting together sacrificing their individual property

14. MICHAEL JACKSON YOUR SONG "MAKE IT A BETTER PLACE TO LIVE FOR YOU AND ME AND ENTIRE HUMAN RACE" WILL BE SUNG AS "BETTER PLACE FOR EVERYONE TODAY AND EVERLASTING TOMORROW".

rights and started eating and praying together. However, Christianity under influence of selfishness became materialistic and made adjustments to suit the PSSS during the 4th century AD onwards. It was subsequently hijacked by the Europeans and twisted the teaching of Jesus to suit the PSSS. It failed to deliver and got divided into several wings each propagating its own ideals. There are people especially Europeans who feel that they have more freedom of speech who are questioning the life style of Jesus. They doubt his accomplice Mary Magdalene. BUT THESE EUROPEANS WHO ARE ROOT CAUSE OF ALL EVILS OF PRESENT SYSTEM have not discovered a method of adopting the principles and the commandments given by GOD. They have during the last 20 centuries have not been able to devise an IDEAL WORLD, which was dreamt by Jesus for which he sacrificed his life. Regarding Islam, Buddhism, Jainism, Hinduism, Zorastrism and all other religions the religious leaders have COMPROMISED with the PSSS. They have built their empires and are not ready to compromise with each other to evolve a common religion. They all know that GOD is one and therefore it is not correct to misguide people, which many time turn into violent conflicts between different religious groups.

Religion is more or less misused by politicians for their selfish goals and religious leaders are puppets in their hands especially in under developed and developing countries.

15.   NO MORE INDUSTRIES WHICH POLLUTE ENVIRONMENT TO THE SLIGHTEST EXTENT.

The PSSS, which, thus evolved then, regularised its activities by adopting commerce, economics, laws to regulate crimes etc. **The laws in the PSSS do not concentrate on prevention of crimes but only contemplates punishment after committing crime.** Person can decide to commit a crime depending upon his capacity to receive the punishment. Meanwhile the victim who has suffered from the crime will not get any relief unless the crime is proved. The kings and despots who were in power under PSSS have committed severe crimes against humanity. We enjoy the great workmanship of huge structures and palaces of these kings and despots. But we are not noticing the pains, suffering, and loss of life of ordinary citizens who were forced to work as slaves by these barbaric kings and despots. Even religions have failed to stop these kings and rulers from acts of barbarism. Instead, even after these kings and despots have been overpowered, we appreciate them by writing on their heroic acts and their patronage for sculpture! The intelligent historians who were in turn their patrons have suppressed their deeds against humanity.

The 20th century democracies have adopted selfishness by adopting competitive markets, multi-national companies and wars that are all crimes against humanity. **Competition creates an atmosphere wherein the survival of fittest alone is adopted.** Wars have destroyed property and life of ordinary citizens. Multi-national companies are adopting methods to boost their

16. WE SHALL WALK. WHY WE SHOULD PUMP CARBON DI OXIDE TO ENVIRONEMNT. NO NECESSITY OF MIGRATION BECAUSE SWS GURANTEES FOOD AT YOUR BIRTH PLACE. YOU SHALL ALWAYS BE NEAR TO YOUR DEAR ONES.

wealth at the cost of humanity. The **market system democracies have increased the gap between have and have-nots.** In the result it is depriving majority of their basis necessities. The Politicians are becoming rich at the cost of poor voters. The elections are fought not for serving the people but to destroy the people. Once in power laws and rules are formed for the detriment of the common man. **People are striving hard to come out this shackle of bureaucracy and democracy.** Selfish deeds are being regularised to save the guilty and corrupt people. There are certain persons and entities that enjoy wealth equal to the budgetary requirement of some of the small developing countries. The PSSS recognises VVIP and VIPs. The divides between have and have-nots is clearly visible not only between persons but also among nations.

All irregularities like adulteration, brothel industry, unlawful activities are committed with the knowledge and support of bureaucracy in majority of the countries.

Money power and political power are the mantra of the day. There is scant respect to human rights. For the sake of money people engage in all kind of evils. **Money is the main theme of every activity.** People are ready to kill others for the sake of power and money. Education and Religion are commercialised. People have forgotten everything in this mad rush for money and power that they are basically animals with rationale. Corruption is the order of the day in all walks of life. The craze for

17. DO NOT WORRY ABOUT TOMORROW. BE HAPPY.

money is so much that there is no limit for its acquisition. The preaching of great saints, martyrs and religious gurus is thrown to winds for the sake of money.

THE PSSS DOES NOT RECOGONISE UNSUCCESSUL people. To achieve success all methods legal and illegal are being adopted. Distrust and fear is fast spreading especially among younger generations. The pressure adopted by the parents on children to achieve SUCCESS is great. It has become a mad, mad world. Cut throat competition, survival of the fittest, recognition of successful, are the basics of PSSS. It has destroyed human values thus depriving poor people who are unsuccessful to slavery.

Majority is deprived of basic necessities and is being wiped off by hunger from the face of this universe. Weapons of mass destruction have been devised to show the power of a nation. Terrorists are killing innocent people to show their presence. Thus the common man has become a soft target for all-powerful persons.

Billions of dollars are spent on defence and research for wasteful purpose. Scant respect is paid to the laws of nature and to maintain its virginity. Forests are being destroyed for commercial purposes. The power mongers have become mad and trying to achieve short-term goals at the cost of humanity.

18.   ONCE SWS IS ESTABLISHED EVERYONE
WILL REMARK HOW FOOLISH PEOPLE WERE
IN PSSS.

DEVELOPMENTAL MODEL OF PSSS

The developmental model of PSSS is short-visioned and short term. The rich people want all the luxuries. They are not bothered about environment. The USA was the first to oppose reduction in green house gases. The PSSS cannot have long term plan say for a century because its model of development does not suit it. It is worried about today. It consumes and ejects mammoth quantity of green house gases. It affects everyone. The US model consumes lot of crude oil to maintain the running spree of its citizens. Thus money power indirectly affects the environment. The US Government cannot think of any model where it can make its citizens consume less crude oil. The developmental indices are based on how much more crude oil is consumed!. The US citizens are not bothered about the damage they cause to the environment because they can only plan for the day. This applies to all developed countries and also to those developing countries that try to copy US models.

The developmental model of PSSS in developing countries is to copy the same from developed countries. Most of the developed countries have limited population. The developing countries and under developed countries which have a big population also copy the model of development of PSSS. E.g., countries like India and China have huge population nearly one third of world's

19.   GOD! WHY 10 COMMANDMENTS. WE HAVE ESTABLISHED SWS. WE HAVE DISCOVERED YOUR IDEA OF HEAVEN ON EARTH.

population. Hence it can utilise its population effectively in manual irrigation without resorting to power generated irrigation. Similarly the Building industry is using heavy machinery instead of manual labour. In the process while it not only creates more unemployment but puts more pressure on limited resources. The developing countries always face shortage of power because of the wrong model of development. In the result more nuclear and thermal projects based on coal are implemented creating serious health and environmental hazards.

In a nutshell the nations of the world are pursuing developmental models which are based on selfishness with least consideration for the future of the Mother Nature.

FLAG OF SWS

The flags of PSSS nations are all coloured majority representing their true nature. Red is the major colour of all PSSS nations. The flag of SWS will be combination of all colours and that it will be pure white representing cleanliness in thought, action, plans and deeds. The white colour represents truthfulness, honesty and fraternity. The flag of SWS will be pure white cloth of pure cotton representing simplicity and equality.

20. SWS WILL TAKE ALL NUCLEAR WEAPONS AND OTHER ILLEGITIMATE ASSETS OF PSSS TO MARS AND DUMP IT. SWS CANNOT KEEP IT ON EARTH. SWS WILL DESTROY ALL LITERATURE CONNECTED TO DISCOVERY AND FORMATION OF NUCLEAR AND OTHER WEAPONS WHICH WERE HARMFUL TO MANKIND AND NATURE SO THAT NOBODY CAN REVIVE SSS.

6
TRACING THE ORIGIN OF PSS

From the ape, originated a creature with thinking powers called MAN. He started his career unlike ape by hunting because he can think well than the ape. It is told that once, while he was hunting, he came across a fiery bush and besides it there was a dead sheep. He immediately went and grasped it. To his utter surprise it was very hot and his fingers penetrated the flesh. In a reflex action his fingers went inside the mouth and thus Man tasted the first mutton. It was wonderful. He started the art of cooking. Then he slowly he thought why should not I grow some crops. He started agriculture. Here he started to settle near his fields and started the first selfish act of earmarking property rights. **The stronger men started to torture the weaker one and thus came into existence slavery.** Then started the race for grabbing more and more land. **Thus came into existence kings and monarchs. Wars were fought between these kings to grab more land.** It went unto the beginning of the 19th century. Of course due to certain bold people like Abraham Lincoln, Mahatma Gandhi the kings and queens were forced to hand over power to democratically elected Governments. However, the remnants of kings and queens were able to still exert influence on governance and the scheme of Democracy in most of the countries is still in par with Kings and queens.

21.   NO MORE AUTOCRATS, BUREACRATS, DIPLOMATS, DICTATORS, KINGS. IN SHORT NO MORE LEADERS. ONLY VOLUNTEERS TO RUN COMMUNITY CENTRES.

Several kings and queens have thrived on the slavery of the downtrodden and like everybody they have all perished. Thus the foundations for the modern PSS is selfishness. Now all the irregularities of land and wealth grabbing has been regularised by way of laws. Regularisation of irregularities was the main achievement of Democracies during the initial years of its birth!

Religions were created to regulate the ill doings of kings and queens. It created a moral barrier for the downtrodden to rise against the rulers and rich. **God was used as an instrument to divert the attention of the poor towards the wrong doings of the power mongers.** This has continued even to this date.

In the 21st century the regularisation of all the wrongs of the past were completed. Each country under the guise of so called democracy has grabbed its share of property of the earth. In the process today 80% of world's population is below the poverty line. **Poverty line here is defined as a person who is deprived of his daily bread, clothing and shelter to which he is rightfully eligible.**

It is pastime for the rich to visit palaces of kings. Everyone wonders at the structural magnificence but nobody ever thinks of the millions and millions of poor people who perished for its construction.

TODAY MAN HAS BECOME SO SELFISH HE DOES NOT FIND TIME TO THINK ABOUT THE FATE OF THE POOR MAN.

22. *ALL WILL WORK FOR EVERYONE AND ENSURE FOOD, SHELTER AND CLOTHING TO EVERYONE.*

7
FAILURE OF PRESENT SYSTEM

The PSS is a failure on more than one count. It only thinks of administering medicine for all illegal activities without devising methods to curb such illegal activities. E.g., it makes laws to punish the person committing rape on women. It does not devise methods of curbing rape because it wants rape to continue because it cannot prevent its powerful members from doing such illegal activities. It's member nation build up huge arsenal of arms to fight its enemy nation without devising method to avoid any such conflict. The UNO has become a puppet in the hands of the powerful nations. VETO power itself is a clear indication of the high handed approach of the selfish nations.

The PSSS has fallen into a vicious circle that it cannot solve its problems but gets entangled in more and more problems, simply because it goal is selfish.

The brothel industry is a clear indicator of the selfish nature of PSS. This damn industry is being recognised as a legal affair in almost all the countries!

Alcohol which destroys so may families especially in under developed and under developing countries is so strong a force that it has become legal entity. The women who are the major sufferers because of brothel and alcohol industry are helpless because the PSS is male dominated

23.   SWS PROVIDES ENOUGH CLOTH TO
WOMEN SO THAT THEY CANNOT SHOW
THEIR BREASTS AND SEXUAL ORGANS
TO PUBLIC. THEY CAN RESERVE IT TO
THEIR HUSBAND AND THEIR CHILDREN
ONLY. OFCOURSE NO BENEFIT IN PUBLIC
DEMONSTRATION BECAUSE SWSPROVIDE
ONLY FOOD AND NOT MONEY.

entity. The selfish male population of PSS is capable of forcing the women folk into slavery throughout the world even in the 21st century.

Apart from this there are so many industries which are thriving in PSS which are detrimental to the cause of the poor people of the world.

THE PSS RECOGNISES ONLY SUCCESSFUL

The PSS recognises only successful. The failed lot will be dragged into slavery. The exam patterns set by PSS forces students to burn the midnight lamp in their studies. In the process they loose their health. The survival of the fittest theory is being adopted and life has become mechanised. Further to pursue education huge investment is required. Thus the doctors son become doctors, the bankers son become bankers and the lawyer's son become lawyer. It is estimated that to become a doctor there is a requirement of investment of Rs.50 lakhs to Rs.1 crore in developing countries. Thus a poor man's son will remain poor.

There is no moksha or emancipation for a poor family. The political leaders and public government servants adopt corrupt practices and collect bribe to become successful. In the process the Government agencies have become helpless to render any assistance to its citizens. The craze to amass wealth by government and semi government servants has become a daily routine. Thus citizens are

24. NO MORE THEFTS, BRIBES, SCANDALS, DACOITY, OCCUPATION AS ONE CANNOT AMASS WEALTH FOR TOMORROW.

being harassed and forced to break the laws to fall into a trap. It is now said that the kings are far better than the present day rulers. The kings had fear of a rival king's attack whereas in the PSS the rulers have no such fear. The UNO is created to stop wars between nations. Bureaucrats and politicians are robbing the public money paid into Government. Huge amount is being swindled on wasteful projects and sometimes actually executing the project. The audit section has also become a partner in such corruption. The fence has started to eat the fields. This is all because of the selfish attitude of the PSS.

Elections are fought only by powerful. Although it is said that democracy is government by the people for the people it is not the case in most of the democratic countries. Only the rich and powerful can fight elections. The poor cannot think of contesting elections because of the huge election expenditure to be incurred. Elections have become business in majority of the democratic countries. You invest in elections and reap its fruits after winning the elections by swindling public money.

WHAT IS THE BASE OF THE PSS?

In the primitive stage of human evolution man was hunting and only the powerful and healthier species could win the battle. Survival theory was the base. It was quite natural in such a situation. In course of time the powerful and wicked could realise that he can control huge territories and make his life successful at the cost of

25. PSSS----- ASSET = LABOUR+CAPITAL+ LAND +NATURAL RESOURCES------OWNER. SWS-------- ASSET = LABOUR+LAND+NATURAL RESOURCES------PUBLIC PROPERTY.

weaker people by adopting them as slaves. The history of kings and despots continued till the 19th century. When people started questioning the authority of the despots by means of revolution, they evolved a method called democracy. But today democracy has become hypocrisy.

The survival theory still persists. In the PSS everyone is asked to run a race by the system and even at the cost of our own brethren we try to win the race. Money has become the only goal of everyone. The instrument of money has made everyone slaves. We are today so much involved in our race that we do not have time to think who devised this race and whether it is a honest race. We do not find time to think of those who fail in the race. ***WE HAVE NO BOTHERATION THAT OUR OWN CHILDREN OR THEIR CHILDREN MAY FAIL IN THE RACE.***

Today we see a beggar and we have no time **whether the system justified in making him a beggar. Tomorrow our own children or their children may become a beggar. We have been so much commercialised and mechanical that we have lost our conscience and the inner voice is suppressed.** We show some pity by a small donation. But we have no time whether we are justified in snatching his daily bread.

It is said that 50% of world's wealth is in the hands of 2% people. **We can put it is this way. 2% people deprive 48% people of their rightful wealth.**

26. NO MORE GEORGE BUSH, ABDUL KALAM,
A.B.VAJAPAYEE.
THERE WILL BE A6WB363, I8DC462, P9LT115

Oh Jesus! Your sacrifice for the poor and the downtrodden has been commercialised by the Christian Church. The Europeans who took control of the Church have devised selfish methods to downgrade your sacrifice. The rich and powerful is ruling the Church. Your sacrifice has been totally neglected by the Church. They have lost their mind and have made adjustments to your sacrifice by their own interpretation. Your teachings were simple. They are making it complicated. You have taught us your prayer. You have said that we can pray only for our daily bread for everyone. Here they deprive a few billion people by withholding their daily bread. **It is farce that those who have daily bread for the next 100 years mechanically utter the prayer you have taught**. They are not ready to accept that your teachings are clearly a method of establishing PURE COMMUNISM. The Church considers the very word COMMUNISM blasphemy. The Church is entirely fallen prey to the rich and the powerful. The Pope has become a person without teeth. He cannot and will not propagate your theory of COMMUNISM because he fears the wrath of rich and powerful. The Church runs a few congregations that distribute Charity. But no one is ready to **REMOVE POVERTY IN ENTIRITY. All the religions have made adjustments in the teachings of their founders to suit the PSS.**

27.   THERE WILL NOT BE CASTES, CREEDS,
RACISM, FANATISM, RELIGIONS.
THERE WILL BE ONLY MEN AND WOMEN.
OFCOURSE ONE RELIGION WITH OUR GOD.

IF YOU POSSESS WEALTH MORE THAN YOUR DAILY BREAD YOU ARE DEPRIVING SOMEBODY ON EARTH OF HIS DAILY BREAD

Man can create wealth but out of his wisdom he cannot produce a single grain. The produce of the earth is all god given out of love for humanity. But we human beings eat and store more than our daily bread thereby depriving somebody on earth of his daily bread. The same is the case for cloth and shelter. I cannot understand the rationale of the present selfish world. It has lost all his senses and creates laws for the protection and recognition of illegal wealth. Here illegal wealth means possessing more than the daily bread.

If there is a fish or creature, which is tastier, but its survival on this earth is essential, the selfish society mindless of the consequences will destroy that fish or creature because it is tastier. The mindless development schemes of the PSSS are destroying the ecological balance of this nature as there is lack of world planning. Each nation out of its selfish aims systematically destroys the ecology. No one is bothered. Every one is worried of amassing illegal wealth and is worried about GDP and senseless Sensex. The richer have been enjoying their life even in their old age using Viagara! Shame on them. Science is being misused. The good purpose for which a Scientist breaks his head to discover something for the good of the Society is being used for selfish ends. The atomic theory was developed to solve the power problem of the world whereas the PSSS has modified it to produce weapons of mass destruction. The PSSS has already fallen into the trap which itself has set.

10
MONEY BENDS LAWS IN PSSS

It is said Laws are made in PSSS for breaking. In most of the developing and under developed countries laws are applicable only for poor. The rich it is said will slap the face of law with currency and law turns its face away from the rich. The profession of advocates is to help the rich to escape the clutch of law. The witnesses are threatened and sadly the evil doer escapes justice. The SWS is such a system where because of equity and freedom the evil doers will face the music. Further the necessity to break law is less because there is no incentive for breaking the law. Under PSSS laws are broken for money or with the help of money. Under SWS money power is almost absent.

WOMEN ARE THE GREATEST VICTIMS IN PSSS

The PSSS thrives on the principle of survival of fittest and the stronger rule the weaker. The women are weaker as compared to men and they are victims in the hands of stronger men. They are victims of repeated cruelty, rape etc. Most of them work as slaves in PSSS. They have been victimised right from day one of PSSS. Female foeticide is an example why parents avoid girl child. A girl is always a third party between her parents and in laws. If luck favours her then she may survive in their in laws house.

28.   SWS DOES NOT REQUIRE OVERNIGHT
SHEDULES, EXCEPT IN HOSPITALS. SWS
ASSURES MINIMUM 8 HOURS OF GOOD
SLEEP WITH FULL OF DREAMS.

But if she becomes a widow at an early age then she will be discarded both by in laws and parents. Her problem will be further aggravated if she has one or more children. Most of such women land in brothels and forced to work as sex workers for the sake of her children and for her survival. Most of the women are denied the privilege of education. The men have always dominated women in PSSS. The PSSS has meted great injustice to women.

In some communities the women is considered as child delivering machine with least voice against her husband against bearing children. Their health conditions are not taken care by their husbands. They are considered as a commodity without any voice to raise against oppression on her. Even married women are raped by their husbands!. Such cruelty is silently suffered by women because they cannot PSSS. If she opposes then she will be deprived of food and her survival becomes difficult.

The SWS will remove all these inequalities. The guarantee of food, clothing and shelter will give women a level playing ground in par with men. Thus the injustice meted to women in PSSS will be wiped off.

RESOURCES OF NATURE BELONG TO ALL CREATURES

The ecological balance of Nature depends on all creatures from micro organisms to macro organisms. Hence the resources of earth cannot be the sole property of human beings alone. This earth is the home for all

29.   NO MORE CHILD ABUSE, CHILD LABOUR. ALL CHILDREN WILL REMAIN AS CHILDREN ONLY. SWS FEED THEM. PARENTS YOU ONLY LOVE THEM.

creatures. Human being has no right to destroy the habitat of co-organisms. The existence and survival of all creatures is essential for maintaining ecological chain. All creatures are important for the delicate balancing of ecology and environment. The PSSS has lost all ethical sense and is busy in destroying the biological chain.

The food grown on earth cannot be considered the sole property of the cultivator because it is supported by nature on which cultivator does not have control. E.g., rain, environmental conditions etc.

The food is definitely given by Almighty to the benefit of all earthizens out of his love for not only mankind but all creatures. Hence food cannot be made a commercial commodity. The produce should be shared by all earthizens equally.

The mining industry is removing the ores and crude oil which was lying in the earth crust for the last so many centuries. If the life of this earth is estimated as one year, human being entered thee scene a week age and in the last one day he has converted this earth into a dustbin. In the process he is digging his own grave.

The PSSS will simplify life style and therefore unnecessary luxuries will be curbed therey resulting on less mining and destruction of Eco-chain. The SWS will scrap all industries which are dangerous to ecology and environment. The global government will be able to plan and make out which is good and which is bad, ofcourse

30. NO MORE COMPETITION, RUSH, TENSION, STRESS. WE WILL SIMPLIFY LIFE. DEFINITON OF LIFE WILL BE REDEFINED. LIVE UPTO YOUR NATURAL DEATH. NO NECESSITY OF HOARDING FOR POSTERITY.

without selfishness. Even food habits will be changed for the sake of maintaining the biological chain.

WHO SHOULD TAKE LEAD IN ESTABLISHING SWS:

A. Ofcourse every earthizen can take lead. However, as a matter of getting personal relief the following persons should definitely take lead to establish SWS to erase the injustice and oppression meted by Kings, queens, rulers, terrorists, rich and powerful.

1. All victims of war.
2. All victims of terrorism. The 9/11 vicitms in particular should take lead to give relief to the souls of persons who died in the gastardly act.
3. All victims of rape, corruption, partition policies of PSSS, border disputes, accidents etc.
4. All political expatiates.
5. All victims of racial discrimination and ethnic cleansing.
6. All green peace activities, NGO's.

All the above persons or their relatives of the above persons should take lead to do justice to their loved ones who died or became physically handicapped in the above causes.

B. BILL GATES, WARREN BUFFET, NARAYANMURTHY AND OTHER LUMINARIES:

31.   SCRIPTURES OF ALL RELIGIONS
   IN PSSS ENDORSE SWS. SWS SHOULD
   ONLY SACRIFICE OUR SELFISHNESS TO
   UNDERSTAND IT.

These persons have risen from nowhere. Now it is time for them to raise every earthizen from everywhere. Their tremendous potentiality can change the face of this earth. They can become immortals if they take lead in establishing SWS. They can change the face of the earth for good. They have the capacity to remove the scars on the face of this earth created by Kings, despots, dictators, power mongers etc.

C. PARENTS OF CHILDREN WHO DIED OR BECAME PHYSICALLY HANDICAPPED IN FIREARMS USED BY FELLOW STUDENTS IN SCHOOL PREMISES

Your child died because of a firing from a firearm of a fellow student of your child. Here the law of PSSS will catch the student who fired against you son. It does not go to the root. The root lies in the malafide intentions of firearm manufacturers and sellers. Why PSSS permits manufacture of firearms and weapons? Now there are no cruel animals, which were there a few decades ago. The weapon manufacturers by this time should have realised this truth and stopped production of firearms. But unfortunately weapon manufacturing is a thriving business. The PSSS does not have the will to close this industry. The victims of this business are poor students who pay with their life. Hence I call upon the parents of children who died in firing in schools to rise against PSSS and strive to establish SWS.

32.   FINALLY WHY LIVE A LIFE AT THE COST
OF LIFE OF ANOTHER HUMAN BEING.
SOFTEN YOUR HEART; THINK RATIONALLY
AND HELP ESTABLISHMENT OF SWS.
LET US MAKE EVERYONE A SUCCESSFUL
HUMAN BEING. LET US LIVE AND MAKE
OTHERS LIVE.

ADVANTAGES OF SWS

1. The abolition of selfishness will remove all evils attached to the world. These include hatred, violence, robbery, dacoity, terrorism, etc.

2. As all are given food, clothing and shelter there will be total welfare of mankind.

3. **All will become equal in all respects and there can be no more slavery or subordination**.

4. The women and children who were hitherto oppressed will be able to lead a life of dignity and happiness. The sins committed by men on women will be wiped off.

5. The main advantage of globalisation will remove boundaries and the hitherto wasteful expenditure by Governments on defence will be removed, thereby reducing the burden on people by way of taxation.

6. Wars will be things of the past. The unimaginable loss suffered by humankind by the acts of a few powerful and selfish leaders in waging wars will be wiped off.

7. There is no FEAR OF TOMARROW. The torture faced by children in running the race of life is avoided. Education will be only for educating others and gaining knowledge. Education will have no relevance to future life.

8. The discoveries and inventions of scientists will only be used for good of mankind and not for its

destruction. It is a shame that PSSS has stockpiled weapons of mass destruction. Scientists will be asked to devise methods which will provide the best food, clothing and shelter without damage to nature.

9. The damage caused to nature by PSSS will be repaired. Unnecessary industries and services, which were damaging to nature, will be closed down. Maximum effort will be utilised to restore nature to its virgin form. The eating habit of mankind will be suitably modified to the extent it does not damage Nature.

10. There will be real independence for everybody and no one can dominate another. Ofcourse as the theme of life will change from selfishness to Love the words like domination, subordination etc will be of little consequence.

11. There will not be oppression on caste and colour basis. All will be equal. There will not be regional imbalances and hence no evils attached to it like war, security etc.

12. The unnecessary pressure exerted by PSSS on the limited resources of nature will be removed. Life will be simplified without requiring luxuries at the cost of nature. E.g., a particular species of fish which is very essential to ecological chain is required to be saved, then as per research scientist analysis its consumption by human beings will be banned. People will eat only

such things, which are essential for their survival without damaging ecological balance.

13. Tension between countries, between people and sometimes within oneself will be all things of the past. People can live a life of total tranquillity.

14. Global planning is the most important advantage of the SWS. In the PSSS a country could decide on its resources and was capable of disposing it with least consideration to environmental effect. E.g., Malaysia is cutting its forests, which was there on earth for the last 100 years. The persons who are cutting are not born when the trees are planted or born. It is unfortunate that the international community could not raise its voice because the system does not provide for it. Malaysia may not be aware that its actions will affect the entire earth. Such is the force of commercialisation that nobody is bothered about environmental hazards of such actions.

15. **There will be global planning and Mother Nature will be protected from all the illegal and mindless activities of PSSS.**

LET US ALL JOIN TOGETHER TO ESTABLISH SWS FOR OUR OWN SURVIVAL.

FINAL QUESTION? WHETHER IT IS POSSIBLE TO ESTABLISH SWS?

Why not. We have thrown away kings, monarchs and dictators. Now only the remnants of those cruel beasts exist in the world. We shall wipe them out! Of course with Mahatma Gandhi's weapon of **non-violence**.

CONCLUSION

Dear reader if you are enlightened that this system is the right one for this earth, then please send your opinion on the same. You are required to indicate whether want to be a active member or a passive member. Active members are required to preach, propagate the theme of SWS and participate in elections. They will be messaiah's in their country. You can enrol yourself as a member by sending your bio-data to the following address: **Ai 1 Km 1 Ferns villa, Majila, Mangalore 575002** Once you send your bio-data you will be enrolled and assigned alpha-numerical mark which will be required for future correspondence. The establishment of a new system will have initial expenses. Kindly donate liberally and all the donations will be listed in the web site and all expenses will also be listed in the web site. Donations may be sent to Bank account No.1289831 of ABN-AMRO Bank, Mangalore.

Thank you

ROYSTON FERNANDES
New Alpha Numerical name : Ai Km 1